ALCHEMIST KIDS CLUB

Adventures in Boron

Book 1

By Cate Spencer & Sarah Henry

D1247972

Illustrations by Dorothy Henry
Cover Design by Justin Benson

A Note from the Producer

When searching for a book series for my 9-year-old daughter, Hazel, I was surprised by the absence of adolescent series based on **STEM** content. A 2017 survey revealed that science was the #1 subject kids wanted to learn more about: 45% of the surveyed kids wanted more science in their lives. After discovering this unmet need in adolescent literature, I decided to create a book series that featured one element of the periodic table in each story. My daughter and her friends love making slime and while I researched recipes, I learned that Boric Acid was the essential element in reducing the stickiness of slime.

The characters were mindfully created to reflect real challenges kids face today, ranging from anxiety, learning disabilities, living on the Autism spectrum, and cultural differences. It was imperative to me that readers are able to see themselves in at least one of the characters.

I believe Joss Whedon says it best "whatever makes you weird is probably your greatest asset." When "freaks and geeks" of the world realize their awesomeness and work together, they make incredible achievements.

– Cate Spencer

This book was designed by Rachel Settle & Sarah Henry with the font Dyslexie. Dyslexie was created by Christian Boer to make reading easier for people with dyslexia. The Alchemist Kids Club is all about inclusion and Saffiyah, Asher, Zach, Pearl, and Xi would have wanted this book to be as accessible as possible. To learn more about Dyslexie, visit dyslexiefont.com.

Table of Contents

Chapter 1 - Saffiyah

The tried and true Calligan family Bronco made its way up the winding path, slowly rolling to a stop in front of their new home. After an hour's drive in a car packed to the brim with the family's belongings, the big, green house was a welcome sight.

Saffiyah surveyed the row of houses surrounding a green expanse of sprawling grass. On the grass, a group of kids ran around in circles at top speed, weaving in and out of each other as they shouted and played. She looked up to the house; *not bad as far as first impressions go*, she thought. Considering how much she moved around, she considered herself an expert on the importance of first impressions.

She wanted to head inside to inspect the house fully before she committed to a more thorough assessment. Though it was not the biggest home they had lived in, it certainly wasn't the smallest either. It was a two-story Victorian that had character and charm, all the way from its wooden shutters to the sunlit porch. With an architect for a dad and a mom whose job caused them to move around a lot, she knew much more about housing styles than most kids her age.

She was most excited to see the basement. Her parents promised she could have the basement all to herself — all except for the smallest corner they reserved for her dad's workspace. She didn't mind sharing; in her eyes, it was a more than fair compromise. Besides, she enjoyed taking a peek over Dad's shoulder every once in a while. It was fun to see what kind of building he was designing that day.

As she got lost in thought, she felt a pang in her stomach; she remembered how she didn't have anyone to share her side of the basement with. No matter how cool of a hideout she turned it into, it wouldn't be much fun without friends. Even after only a short drive, she already missed her friends from New York City. An hour's drive seems worlds away without a driver's license. She was only nine, after all.

She tried to raise her own spirits by reminding herself that she would still see them occasionally and they could always keep in touch with a weekly video call. Even though the city wasn't so far from her new home in Mount Tabor, her parents were always extremely busy and it would be hard to coordinate a time for them to take her to see her friends face to face.

"Saffi, hon, will you help your dad figure out where you want your stuff to go?"

Pulled from her thoughts by the sound of her mom's voice, Saffiyah opened the sliding door of their van, walked towards the rented moving truck,

2

and tried to figure out how to best help her dad sort through the labels on each of the boxes.

Ever since she turned seven, she adopted the role of labeling and sorting their belongings. Her keen eye for organization came in handy, and her parents quickly learned she was best-suited for the task. What could she say? She liked things organized.

She got this organizational spirit, and her dark skin and spiraled hair from her mother. She got her creativity from her dad.

"Hey, kiddo," her dad said as he saw her approach. "Can you remind me which box is which again?"

"The first letter is who the box belongs to, second letter is which room it belongs in."

Pointing to the nearest box, she said, "'SBR:' 'S' for Saffiyah, 'BR' for bedroom. Just 'B' is the basement."

"You got it, Smart Stuff," her dad smiled. "Now let's say you help me unload all of this."

She smiled. "Sure thing, Dad."

The process was like clockwork. The Calligan family was so used to moving at this point that they had an unspoken system. Someone stayed at the truck, handing the boxes to the next person, and that person laid each of the boxes in piles according to where in the house they belonged. Then, once all the boxes were out, they moved the boxes in one room at a time. It minimized how much they bumped into each other, and Saffiyah believed that orderly preparation beforehand made the process significantly easier on everyone.

When they could, they moved into places that came pre-furnished, which helped simplify the moving process. This time, though, they hadn't been so fortunate. They had to order furniture, which was also showing up intermittently today and in the days ahead. The various deliveries added just another piece of the puzzle for Saffiyah to organize.

The one element of the equation she wasn't used to was the kids tumbling about on the community green. She tried to get sidelong glances of the ragtag group every time she brought boxes over to their new yard.

There was a lot to take note of. She noticed that one of the boys was missing half of his right arm, one of the kids had bright purple hair, the other boy was shorter, wore glasses, and always seemed to be talking a mile a minute, and another girl sat perched up in a tree, reading a book with a science-y looking cover. Every so often, she'd look up from her book, scowl if the kids got too close or too loud, and return to reading. Saffiyah wondered why she hung around them if she didn't like the company.

Part of her wanted to investigate, and maybe even

4

join in on this game they were playing. They seemed to be having an awful lot of fun.

She could investigate later. Right now, she had boxes to sort.

While she continued her rhythmic motion from truck to yard, yard to truck, truck to yard, box after box, she thought about just how lucky she was. Not many people with divorced parents got to live with both of them at the same time. Because her mom and dad were still friendly, they shared a home so they could coparent. They wanted Saffiyah to grow up with both parents firmly in her life, never having to choose one over the other.

While the arrangement shifted slightly each time they moved, this house was absolutely perfect for them. It was split almost exactly down the middle, so that each parent could get one side of the house and Saffiyah could go from one side to the other as often as she pleased. In one night, she could get Spanish homework help from Mom, and a moment later, she could get some math tutoring from her dad, all while their parents maintained their independence.

Saffiyah lost count of how many boxes she moved, but before too long, she noticed that the truck was nearly empty. Her mom approached her, put an arm around her shoulder and said "Saffi, hon, why don't you go check out the neighborhood? It's nice out, and there's still plenty of daylight.

You could go see what those kids are up to. How does that sound?"

Saffiyah's chest tightened. Sure, she had been watching the kids for a while, but actually going up and meeting them? That was another matter altogether. The one thing about moving she still hadn't perfected was making new friends. Once they started a conversation, she was okay, but that was the hard part: starting. It made her anxious and she didn't know how to make that anxiety stop.

"Okay, Mom. I guess I could try," she managed to mumble.

"That's my girl," Mom whispered, giving her a kiss on the forehead and a pat on the shoulder.

Saffiyah sighed. Parents could be so embarrassing.

Chapter 2 - Asher

Asher noticed the Bronco as soon as it pulled onto the street. It was Brittany Blue, and he was pretty sure it was the 1974 model, and its condition wasn't half bad, considering its age.

Asher knew he had never seen that car before in-person. Mount Tabor was an extremely small town, after all, and he had a mental index of all the cars in the neighborhood. He wondered what the new car was doing here.

"Asher, focus! Xi's gonna get ya!" his twin brother Zach exclaimed. Asher zapped back into focus as Xi came barreling towards him at lightning speed. He was a hair too slow and Xi caught his right shoulder. He didn't mind. He enjoyed being the lava monster after all.

Even though he was focusing on the game at hand, he couldn't help but notice there was a kid helping unload the boxes. She even looked about his age! Maybe she liked science, too. Science was his absolute favorite and he had loads of facts to share. He made a mental note to ask her soon.

He zoomed around after Zach, sprinting just past Pearl who sat curled up with her book in the tree. In that particular tree, they found that if you climb up about three feet, there's a perfect ledge for sitting. Pearl liked to perch there the most, because she liked the bit of privacy while staying close enough to everyone else. He didn't understand why she didn't want to play. Playing games was so much fun, and he loved zooming around the yard.

He wanted to see how close he could brush past her before she yelled at him. He knew he shouldn't, and he wouldn't admit it aloud, but he enjoyed getting under her skin sometimes. She could be really annoying, so it was only fair to annoy her back, he reasoned.

He wondered how fast he was running. Scientists estimate that cheetahs can run at a top speed of 76 miles per hour, but only for a distance of 1,500 feet, so that would be rather useless in Lava Monster, a game that required some level of endurance. He wondered how fast a cheetah could run a mile. Would they have to run at top speed and take breaks? Could they pace themselves? How could he test...

"Earth to Asher!" Zach called. Asher's face burned red. He hated that phrase. He was already on the Earth, it's not like Earth could come any closer.

"Zach, knock it off!"

"Dude, you stopped in your tracks! What's the deal?"

"I was...I was just thinking!"

"What about?" Xi asked, jumping out of the nearest maple tree. He wondered which of the 128 variations of maple this one was. He couldn't believe he'd never thought to research it before.

"How fast a cheetah could run a mile in. You see, they can go really, really fast, but only for about 1,500 feet, so I was wondering if they could..."

"Pace themselves?" Xi finished. They brought their pointer finger and thumb to their chin and put their hand on their hip in thought. "Hm. I wonder how you could test that."

"Did you guys see the new family moving in?" Zach asked, joining them.

"Yep! I noticed them as soon as they pulled up," Asher announced proudly.

"Did you see that there's a kid our age?" Xi pointed out.

"I wonder what her name is," Asher said, assuming the same thinking position as Xi, crossing his arms and tapping his chin eruditely. That was a new word he just learned. He was eager for a chance to use it in conversation.

"Maybe you could..." Zach began, but Asher didn't hear the rest of the suggestion. He was already zooming over to the girl who was standing there watching them. Her parents were still moving the boxes, but she had stopped to look at them.

She looked rather surprised when he approached her. Maybe even nervous? He couldn't tell. He wasn't the best with facial expressions.

"Hi, I'm Asher!" he exclaimed, "And did you know that the biggest rubber band ball was 9,032 pounds? It was made in Florida."

"Oh, um...no, I didn't. My name is Saffiyah. It's nice to meet you Asher," she replied.

Nice to meet you. So this was going well. Good. He made a mental note to remember to share the rubber ball fact when meeting new people. It seemed to go over really well.

"Come on, come meet my friends, you'll really like them! Also, did you know that there's a record for longest human tunnel traveled by a dog on a skateboard? His name is Otto. He went through 30 people's legs! Isn't that so cool?" he exclaimed.

"Wow, that is pretty neat," she replied, smiling. He put out his hand, she took it, and he half marched, half sprinted her over to his friends. He normally didn't like to touch people, especially strangers, but she seemed a little nervous and he wanted to reassure her that everything was going to be okay. Maybe if she was still nervous, he would pat her on her back. In movies, people did that to reassure other people. He didn't see the appeal, but he wanted to make a good first impression.

"Hey, guys!" he exclaimed as the pair almost barreled into Zach and Xi, "This is my new friend Saffiyah, and I decided she's cool and that she should hang out with us! This is Xi, and this is my brother Zach." Zach extended his left arm to shake Saffiyah's hand kindly.

"And this is Pearl," Asher said pointing to her. "She's really annoying," he added louder than he intended to. Subtlety was never his strong suit.

"I am *not* annoying!" Pearl declared, jumping out of the tree and stomping towards him. "If you want annoying, look in the mirror, Asher," she retorted, snapping her book shut. Asher looked at the title quickly. It was the new Explorer Academy book from Nat Geo he'd been meaning to read. He wanted to ask her if he could borrow it, but something told him this wasn't the best time.

"Hey, guys, chill out," Zach stated, getting in between them. Zach always tried to keep the peace.

Pearl crossed her arms and emitted an obnoxious huff. "I'll cool it when he cools it."

Asher crossed his arms in the same fashion. Keeping up with people's expressions was tough work! "I'm as cool as the largest snowman ever built. It was 113 feet and seven inches! Which is really cool." He knew he was supposed to stay angry and serious, but how could he when he was thinking about such a cool fact?

"No one cares, Asher!" Pearl exclaimed, standing up, putting her book under her arm, and marching off.

Now that really stung. How could you not care about something that amazing?

Chapter 3 - Pearl

Pearl had spent the afternoon at the park reading, but hadn't made it nearly as far in her National Geographic book as she would have liked on account of how loud Zach, Asher, and Xi were being. They kept shouting and wailing and carrying on, and it made it extremely difficult to focus on her Explorers book. She was trying to decipher a code she made based on the book, but it was so hard to focus with all the hollering in the background.

When Asher brought over the new girl and told Saffiyah that she was annoying, that was the last straw. Not only could she not focus on her book, but she was being insulted in front of the new kid! She simply could not stand it any longer.

Not that she much cared what the new girl thought of her anyway. Pearl didn't know anything about her and felt no compulsion to impress her. But Pearl's parents were always on her back about "spending time with kids her age" and "working on her social skills" so they might make a comment that she left early. No matter. She had spent nearly an hour outside with those kids and she thought that should count as enough social interaction for the day.

She really was trying to work on her social skills—her therapist always said it was something she needed to focus on, anyway—but these kids rarely listened to what she had to say and if they wouldn't listen, what was the point?

She walked into the house, closing the door behind her. Her mom was gone on one of her trips again, so the house seemed quiet with just her and her dad. She tried to close the door quietly, but it squeaked, just as it always did.

"Hey, Pearl, how was spending time with your friends?" her dad asked, walking into the entryway. It was almost like he was sitting in the living room waiting for her to enter.

Pearl tried not to get annoyed. She knew he got really lonely, especially when mom was gone. He was always telling her, over and over again, just how fitting Pearl's name was: hard on the outside, but precious to those around her. She didn't really know anyone who found her quite as valuable as her dad.

"There's a new kid moving in up the street," she answered quickly.

"I saw the moving truck. Did you get to talk to her?" Sigh. He always wanted so many details.

"A little bit. I couldn't focus on my book when she came over. Asher was annoying, as usual."

"Well, you could come read in the living room with me, if you want. And maybe you should give the new girl a chance. Maybe she could be a friend you have over here sometimes."

"I'll try." She wasn't sure if she actually would, but knew it was the polite thing to say. Ms. Duncan, her therapist, recommended she try saying "the polite thing."

Whatever that means.

Pearl was trying, but it wasn't easy. Sometimes being polite felt an awful lot like lying. Ms. Duncan said not to think of it like that, but it felt that way just the same. But still, Pearl did like Ms. Duncan, and didn't want to let her down. If the kids in her grade were as smart as Ms. Duncan, maybe then she'd be more inclined to talk with them.

"Excellent. Maybe next time, a good starting point would be asking where she's moving from. That's always a good conversation starter."

"Alright, Dad. I'll try."

Maybe this time, she would.

Chapter 4 - Xi

Xi was surprised just how well this new girl could keep up with them. Not many kids could pick up the subtle nuances of Lava Monster just as quickly as she did. Saffiyah wasn't the fastest out of all of them, but she was strategic and quick minded, which made Xi respect her quite a bit.

As the sun started to set, Saffiyah put her hands on her knees and breathed deeply. She popped back up and said, "Okay, okay, you guys wiped me out. I should probably go help my parents unpack."

"You could really keep up with us! I'm impressed!" Zach complimented her.

"Thanks! I'm not nearly as fast as her, though," Saffiyah said, gesturing at Xi.

Xi's stomach dropped a bit. They didn't tell Saffiyah yet and they didn't know how she would respond.

"Actually," Xi started, "My pronouns are they/them."

Saffiyah blushed and exclaimed, "Oh, I'm sorry! I should have asked. I'm not nearly as fast as them," she corrected herself. Xi smiled. That was easier than they expected.

"And thanks for including me," Saffiyah added.

"Do you think you can come play with us again, Saffiyah?" Asher asked eagerly.

"Yes, I think so! And you can call me Saffi. And maybe once we get settled in, you guys can come have pizza at my house! My parents let me have the basement all to myself. It could be like a clubhouse!"

"A clubhouse? That would be *so* cool!" Asher exclaimed, jumping up and down. He wasn't the best at reading other people's emotions, but he wore his right on his sleeve.

"That sounds great, Saffi," Xi said, smiling. "We'll see you around! I live right over there," they said pointing to the white house, "Zach and Asher live there," they said pointing to the blue house, "And Pearl lives up there," they said, pointing at the yellow house Pearl stomped into about a half an hour ago. "So you can just stop in whenever you want to hang out!"

"That sounds great!" Saffiyah replied. "I'll see you guys later!" she bounced off in the direction of her new house. Xi was really excited about this clubhouse idea.

This could turn out to be an even more awesome summer than they expected.

Chapter 5 - Zach

"Zach! Zach, Zach, Zach!" Asher called out. Zach could hear his brother's feet bouncing off the ground.

"Geez, what's got you acting all crazy?" he asked. He liked his brother's enthusiasm, but sometimes it was a bit much for 9:00am on a summer morning.

Asher was so excited that he didn't even waste time pouting about getting called crazy. That normally annoyed him. *This has to be something really exciting, then*, Zach thought.

"Look! It's a secret note! And it's *coded*!" Asher squealed, pushing the note directly into Zach's face.

"Give me that," Zach said, swiping the note from Asher's hand as Asher waved it excitedly. It took his eyes a second to focus on the no-longer-wiggling paper.

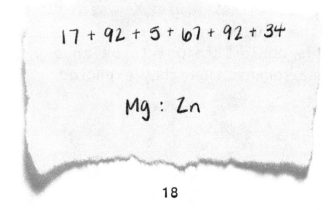

Zach looked up at Asher to see his fists balled up against his cheeks, a bright smile on his lips, and his toes tapping excitedly. Zach had to admit, Asher looked as excited as Zach felt. This was next-level cool.

"I wonder who sent it?" Asher said excitedly.

"Well, I don't recognize the handwriting, so I'm going to assume it's Saffi," Zach replied, handing back the note so he could stroke his chin thoughtfully. This was going to take some real detective work to figure out the puzzle.

"Oh, I bet you're right!" Asher said. "Come on!" Asher tore through the hallway into the kitchen to the drawer where the pens were kept. He riffled through it quickly, found his favorite one, and sat down at the table quickly to start cracking the case.

Zach smiled to himself. This summer was getting more and more exciting every day.

"Now, where to start, where to start..." Asher said, tapping the pen against his teeth.

"Well," Zach started, sitting down beside Asher, "We've got numbers, and then we have letters. Do you think it's two different codes, or do they use the same pattern?"

"Well, all *good* cryptology uses one cohesive code on one problem. So I'm going to assume they're related, because Saffi seems like a good cryptologist," Asher said.

"Good point. Let's start with the letters, then. Where could they be from?"

Asher clicked the pen against his teeth. "Mg, mg, mg...milligram, geometric mean, Madagascar..."

"Wait a second," Zach said, his brain clicking. "Abbreviations, numbers...magnesium. It's the atomic numbers and symbols!"

Asher almost jumped out of his chair. "Yes! Yes, yes, how did I not think of that, I'll go get my periodic table." Asher bolted from the living room, clambered up the stairs, and hurried back down, periodic table poster in hand before Zach could tell him to just look it up on his phone.

Asher spread the poster out on the table and they got to work.

Before too long, they cracked the code.

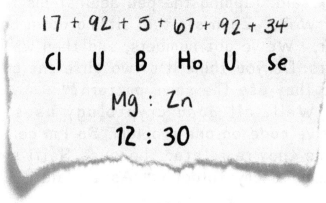

$$17 + 92 + 5 + 67 + 92 + 34$$

Cl U B Ho U Se

Mg : Zn

12 : 30

"Well, now we know who the author is!" Asher said. "You were right, it is Saffi. She's inviting us to the Clubhouse! At 12:30!"

"Yeah!" Zach said, smiling. He was pretty proud of himself that he cracked the code. Sometimes Asher answered questions so quickly that he never got a chance to think of the answer. It was nice being the successful sleuth for once. "Now the biggest dilemma we have is being patient enough to wait until 12:30!"

Asher's face fell like a cartoon character's. "Oh, boy. I'm not so good at that."

Zach laughed.

Chapter 6 - Asher

Asher could feel excitement rising from the tips of his toes to the top of his head. Anyone could come knock on the door to invite you over, but how many people would think to pass out secret notes? Secret *coded* notes! This new girl was turning out to be just as interesting as he had hoped she would be.

At about 12:15, Zach and Asher finished up their grilled cheese and told their mom they were headed out. She told them to go off and have fun. Asher smiled, excited. He had no doubt they would.

"Zach, what do you think the clubhouse is going to look like? What do you think we're going to do there? Do you think we're going to play any games? Do you think we'll play Science Explosion? I love that game. Man, homemade volcanoes never get old..."

"Asher!" Zach exclaimed. "Calm down! We'll find out when we get there. Let's let it be a surprise, okay?"

Asher felt himself deflate a bit as his shoulders slumped. He didn't like being interrupted. "I'm sure whatever it is, it will be really fun," Zach said, trying to coax a smile back onto Asher's face. It worked.

Asher sighed, trying to recollect himself.

Sometimes, he just couldn't help it. When you get excited about something, why keep it in? Being excited is so much fun!

They made it to the green house and went around the side where the land dipped down low enough to make room for a door to access the basement.

After Asher knocked, they both heard the click of a lock unlatching. Asher looked up at Zach, his face bright with excitement. "Go ahead," Zach encouraged. Asher opened the door.

Asher couldn't help but let his mouth fall open when they walked into the room. It was even cooler than he had imagined.

The room had a bunch of mismatched shelves, all covered in books. On top of them, he saw globes, pink Himalayan salt lamps, and trinkets and snow globes from all around the world: Aruba, Peru, China, Egypt, England... from the looks of it, Saffiyah had been *everywhere*.

The walls were covered in tapestries and posters that had world maps, periodic tables, mountains, and trees. There were couches, chairs, ottomans, cushions, and beanbag chairs, which were also all mismatched. It gave the room an eclectic, vibrant feel to it.

"Zach! Zach, look at this!" Asher cried out, sprinting towards a SmartBoard. He quickly snatched one of the markers and started drawing on the screen. His teachers almost never let kids play with the Boards at school. He couldn't believe Saffi had one of her own!

"Asher! Don't touch things without permission!" Zach exclaimed.

Asher quickly put the marker back in its place, erased the marks in one swipe of the eraser and turned bashfully to Saffiyah. He realized his oversight. When he got excited, his manners seemed to fly out the window. He sighed again; he knew he had a lot of work to do before social cues came naturally to him.

"Sorry, Saffi," he said, his head bowed. He peeked up to look at her. "Could I please, maybe, if you'll let me, touch the SmartBoard?"

Saffiyah smiled and chuckled. "Sure thing, Asher, thank you for asking," she replied.

Asher returned to his excited jumping, grabbed a marker, and returned to drawing. Zach turned to Saffiyah. "Thanks for that," he said softly, as though Asher couldn't hear him. He could. Asher prided himself in his excellent hearing. It's not that he couldn't hear, sometimes he just chose not to listen.

At a regular speaking volume, Zach said, "This place looks amazing, Saffiyah! The way you have it set up is so cool."

"Thanks, Zach!"

"So is this whole place yours?" he asked.

"Almost all of it," she said. "See that corner back there?" she asked, pointing to a desk with different drawing implements and graphing tools on it. "That's my dad's office. He's an architect."

"Wow, really? That's so cool. May I see?"

"Sure, you can take a look, but don't touch it. I'm not sure if he has things left in some sort of special position for a drawing. He's really particular about his desk."

"Of course, I won't touch anything."

Zach made his way over to the desk and Asher followed, looking over at a sketch for a skyscraper.

Wow.

Just as he got a good glance, another set of knocks began and he turned to see Saffiyah open the door for Xi.

Chapter 7 - Saffiyah

Saffiyah opened the door to reveal Xi. Saffiyah had been practicing using the right pronouns so that she wouldn't mess up this time. She wanted Xi to feel comfortable in her home and she was determined not to misspeak today.

"Wow, you've got such a neat place, Saffi!" Xi said, taking in the room.

"Thanks, Xi!" Saffi ran over to the corner with the lightswitch. "Watch this."

She shut off the lights and turned on her constellation light at the same time. Everyone gasped as they looked up where a sky full of stars dotted the basement ceiling.

"Wow!" Asher exclaimed. "I can see Draco!"

"All I see is the big dipper" Xi shrugged. "How are you supposed to find all of these different constellations? Just looks like a bunch of stars to me."

"Draco is Latin for Dragon. Greco-Roman legend has it that the goddess Minerva killed Draco and threw him up into the sky. That's why he's still there," Asher explained.

"I suppose if a goddess had to kill me, my body floating up in the sky as stars for all eternity would be pretty neat," Xi shrugged.

"You don't want to be killed by a goddess," Asher suddenly grew dark. "Greek and Roman gods aren't messing around. In their myths, they do some pretty gross stuff to people and demigods they don't like."

"I know, Asher, I obviously don't want to be killed by a goddess."

Saffiyah flicked the lights back on and turned off the constellation light as Pearl walked in.

"You didn't do the knock!" Asher cried.

Pearl shrugged. "I saw you guys walk in, so I knew it was open."

Before Asher could retort, Xi turned to Saffi and asked, "So, what's this secret meeting about?"

Saffiyah had a mischievous smile as she walked over to a table that was covered in a black table cloth.

"What only the best secret meetings are about," she said mysteriously. She whipped away the table cloth to uncover two familiar looking boxes.

"Pizza."

Chapter 8 - Asher

Once everyone settled in with their slice, they each took seats around the coffee table. Asher paused, unsure which one to pick. Ottomans, couches, chairs, beanbags...he had to focus and pick one quick, but he kept getting distracted with fun facts about the Ottoman Empire.

"Asher, come on over!" Xi called. Asher snapped back to his senses and selected a beanbag. That way he would have close access to the table and would be less likely to spill pizza on himself. Food-related clumsiness seemed to be one of his more unfortunate habits.

Between all the sounds of the munching, there wasn't much conversation happening. Asher looked around and heard all the mashing of dough between teeth and he started to get really uneasy. With nothing to block it out, he was starting to experience sensory overload. He had to think of a way to start a conversation fast.

"You know," he started out, more loudly than he meant to, "Everyone thinks pizza is Italian food, but it was actually made by the Greeks first."

"I don't care where it came from, all I

know is that it's delicious!" Xi added. Now that there was conversation, that meant the chewing was less audible because less people were chewing all at the same time. This was going well.

"Hey, Asher, weren't you telling me the other day about the world record for biggest pizza?" Zach prompted.

"Yes! It had a surface area of 13,580.28 feet squared!"

"That sounds like my kind of pizza!" Xi added. Asher was so pleased that his pizza conversation topic was such a hit.

"They even named it! Ottavia: it means "eighth son" which paid homage to the first Roman emperor, Octavian Augustus. Ottavia was gluten free."

"Wow, Asher, you really like records, huh?" Saffiyah said.

"He *never* stops talking about them," Zach emphasized.

"Yeah. It's annoying," Pearl added. Asher sent her a biting glare, but chose to say nothing. It wasn't his fault that she didn't understand how cool his facts were.

Xi chimed in, asking, "What's your favorite record, Asher?"

Asher started to wiggle on the beanbag, trying not to jump up to his feet. Just thinking about this one made him excited. "A couple years ago, there was a record set for the biggest slime ever by weight. 13,820 pounds. Can you *imagine* that much slime?"

"Wow. I can't even picture what that much slime would look like!" Xi exclaimed.

"I wonder how much space that much slime would take up. I imagine they needed a pretty large space to hold all of it," Pearl wondered aloud.

"Whenever Asher talks about records, I can't stop thinking about how cool it would be to set our own. I bet that would be one of the coolest things in the world," Zach said.

"Well, why couldn't we?" Saffiyah asked curiously. "Slime's pretty easy to make, isn't it?"

"Yes!" Asher cried. "It's only a few ingredients!"

"But there's already a slime record. We'd need to come up with a new way to make a record. What measurement could we use?" Pearl asked.

"I think you answered that for yourself, Pearl!" Xi chimed in. "You were talking about how much space it could take up, so..."

"Volume!" Pearl finished their sentence.

"Is there a record for that, Asher?" Saffi asked.

"Let me look!" He jumped up from his seat, unable to contain his excitement anymore. He was halfway to the SmartBoard before he remembered *ask permission* in his brother's voice. He turned. "May I?"

"Sure you can," Saffiyah said smiling.

Asher excitedly made it the rest of the way to the Board. He put in a quick Google search, pulling up the Guiness World Record site. He knew it extremely well, because he was always checking for the latest record. He was almost embarrassed to say he didn't already know if there was a volumetric slime record off the top of his head.

After searching, he knew why he didn't know about it; *it's because it's never been done*.

"There's not a record!" Asher cried, turning to his friends. "Guys, we could do this! We could set a world record!"

The room jumped into an explosion of excited conversation, everyone talking about just how cool it would be, how they would go about doing it, how to create the best slime recipe, and wondering what the process would be like. Asher looked over and saw that even Pearl was joining in on the chatter. She even had a smile on her face.

Chapter 9 - Saffiyah

Wow! A world record! Saffiyah thought she would suggest it, but she never thought that they would be just as ambitious as her to try it!

But how impossible was it, really? Practically everyone she knew loved making slime and there had to be a step by step process online outlining how you can go about making the record attempt official. All it would take was patience, planning, and a spark of inspiration. She could manage that.

It was only three things, after all.

"So, Asher," Saffiyah started, too excited to eat her pizza, "How do you set a world record? Do you know what the process is like?"

"Does he know what the process is like? He practically has it memorized!" Zach said.

"Okay, okay, so first of all, there's three different routes: account managed services, priority services, or standard applications. Because we're not a business, we don't go with account managed, and because we don't have $800 lying around, we go with the standard application."

"$800?" Xi cried out. "Why is it $800? That's *so much money!*"

"Well when you do priority, your application

jumps the line and gets processed in five business days. *Five*. That's *so* fast! It's kind of like cutting in line, but if you have money, I guess you're allowed to do that," Asher said, rolling his eyes.

"Asher, you know that option is for people with a certain date for an event that involves hundreds of people. You can't exactly wait around twelve weeks for it to be processed in standard applications," Zach added.

"Twelve weeks?" Xi cried out again. "That's *so long!*"

"I *know*," Asher said emphatically. "I'm so impatient!"

"Wait," Saffiyah said, turning to Zach, "How do you know how long it takes?"

Zach shrugged."Asher talks about it so much, I guess I have it memorized too,"

"Okay, Asher, get to the point, what's next?" Pearl interjected.

"So you make an account online, you find the record you want to attempt – or in our case, we select the 'apply for a new record' title – you complete the form, wait *forever* for it to be reviewed, and then you get your instructions on next steps!"

"It *can't* be that simple," Pearl said in disbelief.

"Well, it's not simple. There's lots of paperwork. But there's a right way to do it, and it's all laid out on their website."

"If it were simple, more people would break records," Pearl reasoned.

"But lots of people *do* break records. There were hundreds of records broken just last year. There are over 4,000 records that Guinness tracks. And one of them could be ours!" Asher cried. "The reason it takes so long is because they have a bunch of applications to go through," Asher said, "And then they need to help you pick a date, send out a representative...all of that boring stuff."

"I don't think it sounds boring," Saffiyah countered. "I really like all these rules and regulations. It makes everything really organized and easy to follow."

"Well, you can do all that," Xi said, "I want to make *slime*."

The room erupted into people agreeing, talking about their favorite slime recipe, wondering when they could get started, and talking about how amazing it would be if they made a record.

"Guys, if we're going to make this happen, we've got to have a cool name," Zach said.

"A name? For what?" Pearl asked.

"You know, for like, a club. We should start one! We're all really into science, so we could have some sort of science club. The Kids'

Science Club," Zach announced, his one hand outstretched as though he was brandishing the name.

"Eh, we can do better," Xi said. "I like where you're going, but it needs a bit more zing."

"What about...." Asher mumbled, as though in the middle of thought. "What about, what about, what about....the Alchemist Kids Club!"

"The what now?" Zach asked.

"Of course! Alchemy!" Saffi chimed in. "It's one of the earliest recorded attempts at chemistry. Philosophy scholars thought they could purify different metals and even turn lead into gold."

"That does sound pretty cool, not gonna lie," Xi affirms.

"Motion to make the first motion of the first Alchemist Kids Club meeting to make the name of the club the Alchemist Kids Club?" Zach asked.

"That was a bit hard to follow, but motion accepted!" Pearl responded, laughing.

"Seconded!" Asher exclaimed.

"All those in favor of the motion to begin the Alchemist Kids Club say aye!"

"Aye!" they all shouted, laughing and smiling.

Saffiyah looked around the room and all the shining faces inside it and realized that the kids in the room were just as mismatched as the furniture. It was a perfect fit.

Chapter 10 – Asher

It wouldn't be that hard after all, Asher thought to himself. Fluffy slime was a simple chemistry experiment: how much borax, how much shaving cream, how much saline solution to get the maximum fluffiness? Surely they all knew that.

"So how much borax do your parents have?" he directed at Saffiyah.

The murmuring of the room hushed into a muted silence. Everyone was looking at him rather strangely. He didn't know why, since it was a rather simple question.

"Um...none?"

"None? Really? Well I guess we'll have to get some," Asher surmised.

"Asher, why do we need borax?" Xi chimed in.

"Uh, duh, you need borax for fluffy slime."
Crickets.

Hm. Perhaps they didn't all know that. He supposed it wouldn't be *too* hard to teach them. But only if they were good listeners.

Asher let out a big sigh and sucked in a large gulp of air before diving in. "There's four important ingredients to fluffy slime: glue, borax, contact solution, and shaving cream. The shaving cream is what makes it fluffy, and the glue, contact solution, and borax combination is what makes it slime. The boron molecules become an ion when added to the liquid contact solution, and when that's paired with the glue, those ions are what make the glue stretchier, creating that slime texture."

They all stared at him blankly for a moment. Uh oh. Maybe they didn't understand him. But he didn't know how to explain it any more simply...

"Got it," Saffiyah said, nodding her head.

"Yeah, that makes sense," Zach agreed.

Oh good, they *did* get it.

"Wait," Xi interjected, "I get all that, but why would Saffiyah just have boron lying around? Isn't that one of the elements?"

"Number 5 on the Periodic Table," Pearl nodded.

"Pearl's right," Asher agreed. "But I didn't ask if she had bor*on*. I asked if she had bor*ax*. It's a common household cleaner, normally used for laundry and stuff like that."

"Nope, sorry, we just use regular liquid detergent as far as I know," Saffiyah shrugged.

"So where can we get this boron...I mean, borax?" Zach asked.

"Well, if you want to go straight to the source, the most productive mines are found in California and Turkey, but the one that I think is the most interesting is the Rio Tinto Borax mine in Chile," Asher said excitedly.

"Did you say Chile?" Xi asked incredulously.

"Yep, Chile!" Asher answered. "Why?"

"My brother lives in Chile!" they exclaimed. "Maybe we could talk to him to find out more about boron and how it's mined! I could set up a video call!"

Everyone got really excited about that. Wow! Imagine getting to talk to someone who's actually seen boron get mined! How cool would that be?

"What's the point of that?" Pearl intruded. Asher looked at her and glared daggers. What's her problem? "Shouldn't we be focusing on getting the right recipe? Why would the Guiness World Record people care if we know where it's mined?"

"I don't know, Pearl. I just thought it might be cool," Xi shrugged. They were clearly deflated.

"Well, learning about where it's from could help us understand the process and make us appreciate the science behind it," Saffiyah answered. "And wouldn't that make for a more interesting story to cover? Maybe it could help our proposal get picked." Xi's face seemed to lighten a bit and they smiled at Saffiyah.

"Yeah, Pearl," Asher added, standing and putting his hands on his hips. "Knowing the science behind it is probably more important than actually doing the experiment. Because without the science, it's not even an experiment!" Pearl stood up and put her hands on her hips, too. Asher had to look up because she was quite a few inches taller than him. He gulped. "Well, *Asher, you* might care about the science, but the *record* representative doesn't. They just want to see *results*."

"But how can you *get* a good result without doing the *research* behind what *makes* a good result?" Asher retorted, getting louder and standing up on his tippy toes just to try and match her height.

Pearl growled and huffed loudly, turning on her heels and marching out the door, slamming it behind her.

The room was silent. Asher tried to measure the room's reaction, but it just sounded like silence to him.

"Asher," Zach started, "Try to lay off Pearl a bit."

Asher turned quickly. "But Zach, she's so *annoying*," he said, drawing out the word as long as he could just so he could make him understand.

"You know she's a bit different than us,

Asher," Xi chimed in, "She's got ODD and she doesn't mean to be annoying or rude. You know she's been working on it with her therapist. Try to meet her halfway."

"ODD?" Saffiyah asked.

"Oppositional Defiant Disorder," Zach explained. "She gets annoyed a little more easily and is *very* vocal when she doesn't agree with the way someone wants something done. Her dad told me about it. She used to have an even harder time, but therapy's been really helpful for her."

"Well, *I* still think she's annoying," Asher said, crossing his arms.

"Asher, you of all people should be a little sympathetic. You know that you struggle with relating to people, too," Xi mentioned.

Asher hung his head. He supposed they were right. Being on the autism spectrum always did make it difficult to understand the way other people operated socially. Numbers and facts and science just made more sense than people did.

"You're right," Asher admitted. "I just hate apologizing."

Zach stood up. "I'll go help you out, buddy. Let's go together."

Chapter 11 - Xi

On the day of the call with Xi's brother, Xi got to Saffiyah's place early to get the video chat set up, but since everyone else was so excited about the call, they were already there by the time Xi arrived.

"Okay, okay. You can both be right. Asher, you're right, we won't need very many ingredients, which is a good thing. Pearl, you're also right, we'll probably want to test a few different variations to find the recipe we can best scale up," Zach was explaining as they opened the door. It looked like the apology only lasted so long.

"I bet you don't even know what borax is," Pearl said, smirking.

"Of course I do!" Asher retorted back. "Borax is sodium borate, though it could also be called sodium tetraborate because of how the compound is formed. It's a salt, and one of many boron compounds we use every day. Boron is atomic number five, and its chemical symbol is B. It's relatively rare as far as elements go, but because it is so water-soluble, it shows up as salts. Boron is special because boron is an activator. When you use it as borax in slime, for instance, it's what changes what would be putty into slime. It's what gives it connectivity and stretchiness. It's what makes it slime. Without borax, slime wouldn't be slime."

"Hey, chill out, you two," Zach said, getting in between Pearl and Asher. "We know you're both super smart and know all these facts, but now we're going to call Xi's brother so we can learn more. And I'm sure there are facts we will *all* learn about *together* because no one here knows *everything*." This shut Pearl and Asher up for a moment. Even they had to admit that they didn't know *everything*.

Xi walked over to Saffiyah so they could set up the Board. From over their shoulder, Xi heard Asher sigh and say, "I'm sorry, Pearl." If their ears didn't deceive them, they thought they may have heard an apology from Pearl as well, which was certainly progress.

"Alright, Saffi, how do we set this thing up?" Xi asked Saffiyah, gesturing to the SmartBoard.

"Well, I'm already logged in to the account, so do you have your brother's username?" Saffiyah asked. Xi nodded. "Alright, so we'll just add it here, we'll send him a message, and we'll see when he accepts the request!"

Xi followed through with the instructions and almost as soon as they finished sending the invite, Hector accepted it from his end. It looked like he was excited about the call, too!

Xi started a call and almost momentarily, Hector's face appeared on the big screen.

"Hey, there, Xi!" Hector exclaimed. "It's been so long! I miss you, bud."

"I miss you too, Hector," Xi returned, beaming. "This is Zach, this is Asher, this is Pearl, and this is Saffi. They're all really excited to talk to you, too."

Everyone in the room shouted their hello, smiling and waving. Asher was even bouncing up and down, he was so excited.

"Alright, so what do you want to know, you guys?" Hector asked. "I love Chile and love talking about it, so you'll have to stop me from talking too much!"

"Give us a bit of background on the country, Hector!" Xi said.

"Alright, will do. So the Spanish claim they settled Santiago, but people have been living here for thousands of years, which is way before the Spanish even knew this gorgeous place existed. Originally, it was probably occupied by nomadic groups of people, hunter-gatherers traveling between the coast and the mountains as early as 3000 BC. They probably hunted guanacos, which are a bit like a llama.

"The first known establishment of human settlement in the Americas was known as Monte

44

Verde, and it was in Chile! They likely farmed potatoes, maize, and beans, and began domesticating animals. The villages in the area would have been under Inca reign later, in the fifteenth and sixteenth century, and the Inca Trail likely came to a crossroads here, with a route heading west to the coast at Valparaiso and other routes heading both north and south.

"Of course, if you read a textbook, it'll probably say Santiago was founded by the Spanish explorer and conquistador Pedro de Valdivia in 1541 in the service of King Carlos the First. The city was named Santiago as a result of the Latin word for Saint James. The grid layout of the city was designed by master builder Pedro de Gamboa; many of the old cathedrals in the old city are hundreds of years old. The Church of San Francisco, for instance, was built beginning in 1572 — meaning it's nearly 450 years old.

"But anyway, moving on from the history lesson and on to what you are all more interested in," he went on, "Chile has a ton of different industries going for it. Mining - including copper, lithium, boron, and other minerals — fishing and processing — some of the largest fisheries are in Chile — and metal manufacturing like iron and steel compounds, cement, textiles. Chile's making a strong move toward becoming a really competitive country in the global market, and I'm so glad people are starting to see what a beautiful, wonderful country it is."

"Ooh," Asher interjected, "Tell us about Salar de Surire!" Xi shot him a look for interrupting.

Hector continued, "As you likely already know — or at least, Asher already knows —" they all laughed, "Salar de Surire is a world-class salt flat and national monument, home to borate and sodium sulfate deposits. Watching the salt whip across the salt flats, I promise, is some of the best views you'll ever get."

"I've always wanted to see the altiplano!" Asher cried out. He looked around the group. "That means high plain. The Parinacota volcano is more than 20,000 feet in elevation, despite not being that far from the coast, all things considered."

Hector laughed. "Wow, Asher, you really did your homework! The salt flats themselves are at roughly 13,000 feet above sea level. Up there, there's also the Agua Calientes and the hot thermal pools. You'll see flamingos and vicunas, llamas, rheas, rabbits, and more spectacular mountains and volcanoes, 360-degree vistas—I wish you guys could come and see it for yourselves!"

"Hey, maybe one day we can!" Xi said. "Would you let us stay with you?"

"For you? Anything, hermanite," Hector winked. Xi smiled broadly. They couldn't imagine how amazing it would be to bring their friends to Chile to meet their brother!

"So, Hector," Saffiyah began, "What makes boron special? I'm sure it's not only mined for slime, so where does all that boron end up?"

"You're right, Saffi, it's not all mined for slime," Hector laughed along with her. "Boron's actually been used for a whole bunch of things for thousands of years, and it's got a wide range of uses today. It's put in glass and ceramics because of its resistance to thermal shock, which is when a sudden change in temperature causes the glass to break."

"My dad did that once," Pearl added. "He put hot soup in a bowl and put it in the fridge. Glass got everywhere. I guess it didn't have boron added to it!"

"You're right, Pearl, it probably didn't," Hector nodded. "It's also used in fiberglass, cleaning products like detergent, filaments for lightweight but high-strength material for aerospace structures, shields for nuclear reactors, and it can be used to make a neodymium magnet, which is the strongest type of permanent magnet. It's used in MRI machines, CD and DVD players, headphones, speakers, cell phones..."

"Wow. That's so many things!" Zach exclaimed.

"Oh, and I forgot, you can also use it in insecticide when you've got some troublesome fleas, ants, or cockroaches."

"Blegh," Pearl shivered.

"You said it, Pearl," Hector agreed. "I think they're really gross, too."

"Thanks so much for talking with us, Hector!" Xi said.

"Yeah! Yeah, this was so awesome!" Asher said, bouncing again.

"No problem, you guys!" Hector said. "This was fun! Xi, you'll have to keep me updated on the project. Let me know how this goes!"

"Will do, Hector," Xi smiled.

Chapter 12 - Zach

"Earth to Asher," Zach intoned. He knew Asher hated that phrase, but he used it anyway.

"What?" Asher answered. He sounded annoyed, but it at least got him to answer.

"What's the next step?"

Saffiyah cut in. "It's all on the board, Zach."

Oh. Duh. He forgot.

"Sorry," he laughed. "But Asher, you still have to pay attention."

"Ugh, you're not my mom."

"That would be sort of weird, since we do have the same mom," Zach tried to say with humor. Zach looked up at the SmartBoard to see what the next step was. They had each done a bunch of research to come up with the recipe they thought had the best chance at working. Now they were trying each of them out to see which one had the most success. They were on the last one and with some many different steps for so many different recipes, he was having a tough time remembering which step was which.

Each recipe had slightly different properties, depending on the ratios they each had of borax, glue, saline solution, and shaving cream. Some of them were stretchier, others were fluffier, some of them were stickier, which Asher was very vocal about hating. He didn't like the texture and frequently mentioned how uncomfortable it made him.

The table was covered in packages of food coloring and even watercolor paints, so they could test different colors to see how the slime looked in different shades. There was no harm in considering aesthetics in their slime, after all.

They finished up the last step and Saffiyah worked at kneading it correctly so all the ingredients meshed together properly. She held it up for them all to examine.

"Too sticky!" Asher shouted out.

"I knew you would say that," Saffiyah laughed.

"I'm really digging the color, though," Xi added, turning to Saffiyah, "Purple was a nice touch."

"I agree," Zach confirmed.

He marched over to the board, scrolling through each of the slides of recipes. He picked up the red marker and added the notes "too sticky, but good color," to this recipe. He went through all five recipes and added his thoughts to them as well, ranking each of them based on how well he thought each of them performed.

He put down the red marker and passed the green one to Saffiyah. "Here, add your thoughts. Let's pick our favorites and maybe suggest some changes to make each of them more effective."

"That's a great idea!" Saffiyah said, accepting the marker from him. She added her notes and passed a new color to Pearl and they each took turns until all the slides were covered in notes.

Chapter 13 - Xi

As they finished up adding their notes to the board, Xi returned the marker to its tray and stepped back with the others to investigate their work and compare notes.

Saffiyah's recipe was universally "too sticky." Asher continuously explained how good slime should be like a lighter play-doh: easy to knead and shape and form, but without the same weight or texture. Weight was going to be one of their main focuses; since they were going for a larger volume, they wanted a lighter, fluffier mixture, otherwise the weight would be difficult to handle and support.

On the page for general notes, he had written, "Key is shaving cream foam, a mix of 1 tsp borax per cup of water, using enough glue but not too much, and a little bit of contact solution."

"Well," Xi started, "What are we thinking, fam?"

"I liked Asher's," Saffiyah offered.

"It was a bit too heavy, though," Pearl shrugged.

"I thought yours was pretty good, Xi," Zach added.

"I don't know that it held its shape

very well," Asher countered.

"Well, now that I think about it, there were good and bad parts of all the recipes," Saffiyah thought aloud.

"So what are we supposed to do then?" Pearl asked.

Xi tapped their foot and assumed a thinking position.

Lightbulb.

"Guys, what if we combined the best parts of all the recipes into one perfect recipe?" they offered.

"That's a great idea, Xi!" Zach said.

"That seems like the best way to take the best of each while addressing any issues," Pearl agreed.

"I'll write it, I'll write it!" Asher exclaimed, running to the board. *He really does love that board,* Xi thought to themself.

They discussed and debated but finally, after much erasing and rewriting, the new slide read as follows:

- Squeeze a bottle of glue into a medium-sized bowl (preferably at least a quart-sized bowl).

- Stir in ¾ cup shaving foam for a medium

53

amount of fluffiness.

· Add 5-10 squirts of contact solution; the salinity helps reduce the stickiness.

· Add a few drops (or more) of food coloring to get the color desired.

· In a separate container mix 1 tsp into 1 cup of water.

· Add the water-borax mixture into the bowl slowly, helping the mixture clump.

· Knead the slime into the consistency desired.

· He added a section titled "Troubleshooting tips:"

· If too sticky, add more saline.

· If not fluffy enough, add more shaving foam.

· If it needs more color, add more food coloring.

· If you don't like the scent, knead in a few drops of an essential oil.

Asher turned from the board to see the rest of the group looking at him.

"Thank goodness. We finally got something," Zach sighed and smiled.

"Well? How about we give it a go?" Xi asked.

They all murmured in agreement as they flocked to the table to make what would hopefully be the perfect batch. They each did a different part, all speaking in hushed tones as though they were doing a delicate experiment. They each held their breath as they waited to see if this would be the recipe they'd need to make the record.

Xi kneaded the mixture, a bit of sweat forming on their brow. Everyone was watching and they didn't want to mess this up.

Finally, they lifted the mixture delicately from the bowl and gave it a small squeeze. Everyone gasped as it took the shape perfectly. Xi bounced the slime from hand to hand and said, "Guys, it's the perfect lightness and consistency!" They passed it over to Saffiyah and she passed it to Zach and they each got a turn marveling over it.

This is it! Xi thought. This was what could give them a batch worthy of a record!

Saffiyah nodded. "Exactly. Now we just have to figure out the exact recipe and how to scale up."

Looking around the room, Zach nodded in agreement. "I like it. But look at how many bowls and measuring cups and everything else we already have scattered around this room—let's get this stuff cleaned up and *then* we can talk about scaling up."

Saffiyah shot him a grin. "Thanks, Zach. I really wasn't looking forward to cleaning this all up after you four left." She smirked. "I know you were each planning on leaving me with all the work!"

Chapter 14 – Asher

After a week of working and perfecting and hemming and hawing, the team felt as though they really had the recipe that would blow away the representative and give them a winning shot at securing a record. They couldn't believe how close they felt to achieving this dream. It seemed like only yesterday when they had the crazy idea to make a record. Saffi moved in only two months ago, but none of them could imagine the group without her.

Everything seemed perfect.

Pearl, Asher, Xi, and Zach were all sitting outside in the shade. Zach tossed a ball to himself, Pearl read, Asher paced, and Xi laid down in the grass, looking up at the sky.

"Alright, so we've got the recipe," Xi said. "Now what?"

Asher felt that it was an obvious answer. Didn't they listen to him explain the process the first time?

"So we've got the idea," Asher took a deep breath, trying to remain calm while answering, "Now we've got to do all the paperwork and apply."

"And there were three options, right?" Xi asked.

Before Asher could answer, Pearl said,

"Yeah, don't you remember? And we would do a standard application."

Asher looked at Pearl and breathed a sigh of relief. Thank goodness *someone* remembered!

"Right, right, got it," Xi said, nodding their head. "Now I remember. So we basically fill out all the paperwork, send it on in, and wait?"

"Twelve weeks," Zach confirmed, nodding.

"Ugh, that's *so* long," Xi said, motioning for Zach to toss them the ball. He tossed it their way.

"Yes, but it makes sense," Asher said. "They have a lot of applications to go through, and we'd rather they really take the time to consider ours instead of tossing it in the garbage after looking at it for a few seconds. But I still can't wait!"

"Well, it's not like we've got a reason to rush," Zach shrugged. "We can take the time to be extra prepared so we'll be sure to have the best chance at setting a record that's tough to beat."

"I guess you're right," Asher nodded. "I'm just so impatient! I'm so excited that I can barely contain it!"

"Hey, look, it's Saffiyah!" Zach said, pointing over to Saffiyah who was walking over to them.

"She doesn't look very happy. I wonder what's up?" Xi wondered, tossing the ball back to Zach.

Asher squinted, trying to see what they were seeing. He was wondering how they could tell that she was sad, especially from so far away. Her lips were turned down and her eyes were looking down too. She was moving pretty slowly, instead of with her usual bounce. Maybe that's how they could tell. But maybe she was just tired? It didn't seem like they had enough evidence to support their claim.

"Hey, Saffi, what's up?" Xi said, trying to ease into the conversation.

"Oh, um..." Saffiyah said, looking down at the ground. She was twisting the ball of her foot into the dirt, making a small circle. Asher figured that maybe that was her method of stalling. He also put things off that made him uncomfortable, so he understood where she was coming from.

But Asher was also dying to know what she was upset about so he piped up, "Tell us what's wrong, Saffi."

She looked up at him and let out a big sigh. "You guys might have to break the record without me."

The whole group gasped. "Wait, Saffiyah, why would we have to do that? We can't do it without you!" Xi said emphatically.

"I might be moving away before school starts.

My mom might finish out her contract and there may not be more work for her here." She sniffled, and Asher could see tears welling up in her eyes.

Asher didn't know what to do. He didn't like it when people were upset, especially when they cried. It made him uncomfortable and he didn't know how to make them stop.

"Oh, Saffi..." Xi said, their voice soft.

"You can't move! You can't! We need you, don't your parents understand that? You can't leave! It's not fair!" Pearl said, her fists balled tightly. Asher looked over at Pearl. He had never seen her look that frustrated, not even when he started bugging her.

Chapter 15 - Pearl

She couldn't believe it. This was unacceptable. She was finally happy. All the pieces were finally fitting together. All of them. She had a group of friends who didn't make her annoyed, they had a plan to do something really amazing, and now it might not happen. "I hate this. I hate it. This is not fair."

"Not everything is fair, Pearl," Zach said quietly. Pearl knew that, of course she knew that, but why did it have to be unfair right now?

Asher walked forward towards Saffi slowly, as though he was trying to figure out what to do when he got there. He lifted his hand to pat Saffi on the shoulder. "It's okay to cry. My therapist tells me that all the time."

And Saffi did. And they all sat around and let her cry.

Pearl tried to control her anger. She balled her fists beside her and counted to ten. Maybe it wasn't about her right now. But it was hard to focus on how someone else was feeling when what she was feeling felt so overpowering.

Maybe it was okay for Saffi to cry. She knew that crying was a sign of weakness biologically speaking, so the fact that she was letting down her guard must mean she really trusted them.

Saffi started to wipe her face after a while,

sniffing, and taking a few calming breaths. Pearl wasn't always the most comfortable with showing all those emotions, but Ms. Duncan was always encouraging her to try and relate to people and to express her emotions with her friends, so it was only fair that she encouraged her friends to express their emotions as well.

"So," Pearl started after giving Saffi some time to let out her feelings, "What are we going to do? We have to do something. We can't just let this happen."

"Well, we can't do this without Saffiyah. That's just not an option," Zach said firmly.

"Maybe if she moves, we could work long distance together?" Xi offered.

"No. She *has* to be here," Pearl insisted. Didn't Xi know long distance would never be the same?

"I do think it would fall apart almost immediately without her," Zach said. "She's the only one who can keep us organized and on track!"

That made Saffi smile and laugh.

"You're right, Zach," Xi agreed. "So how do we make sure we do this before she's gone?"

They all knew the answer. But it was impossible, so why even bring it up?

"Well...." Asher started, filling the silence. "We would have to do priority services."

"But we can't afford it," Pearl said. "None of us have anything close to $800." Why would

Asher bring that up? He was just getting Saffi's hopes up for nothing. Pearl was trying to keep things realistic. No sense in wasting time on dreams. It made more sense to be practical.

"What if we could earn $800?" Xi offered.

"How would we earn $800?" Pearl snapped. "None of us are old enough to get jobs." Why did they keep bringing up ideas that would never work?

"Well, I've read about a bunch of different groups that raised money for charity during the record breaking attempt," Asher said.

"What if we were able to raise money selling the slime?" Xi asked.

"How would we sell slime that we didn't even make yet?" Pearl asked. "And how would we afford to make all that slime anyway?"

Zach chimed in at that. "Asher, didn't you say something about how one of the slime records sold the slime afterward as a charity fundraiser?"

"Yeah. When Maddie Rae set the record in 2017 for largest slime, she made 13,820 pounds of slime. They sold most of the slime afterward as a fundraiser for hurricane relief funds."

Pearl was still unimpressed. "Wasn't she already a YouTube celebrity, though? There's no way the Guinness Book of World Records is going to care about us and our attempt." She just felt like everything was spiraling. She didn't know how to find a way out of this. She tried to take a breath and consider the ideas before she shot them down. But she just wanted a perfect solution and wanted Saffi not to move away at all.

"What about pre-sales?" Saffi asked. "I've heard of people doing crowdfunding for different projects to raise money for different products that need startup money to make the things they're selling. What if we did something like that?"

"But why would they buy our slime? What makes our slime different from anything else they could buy or make themselves? Why would they wait to support ours?" Pearl asked. She didn't understand why they didn't see all the flaws in their plans.

Chapter 16 - Zach

Why would people want to come? Why would they give money for slime when they can buy it for way cheaper at the store? They went back and forth, Asher saying how people should care about records because records are important in and of themselves, Pearl explaining that not everyone's as obsessed as Asher, Xi wondering what they could offer as incentive, Zach pointing out each option's challenges, and Saffi wondering how they could afford to cover the expenses.

They couldn't seem to agree on anything.

Finally, Xi broke the chatter. "Well. We've got nothing. How about we each talk to our parents, see if they might be able to help us think of something, and then meet back here tomorrow? We're just going in circles at this point."

The group admitted that Xi's suggestion sounded like the only reasonable option at this point. After making a plan to meet the next day, they each headed home.

"Zach?" Asher said quietly as they walked home.

"What is it, buddy?"

"We're going to figure out a solution, right? I mean, we have to. We're getting so close. I can almost taste it."

Zach almost wished he could give his brother a hug. But he understood that Asher didn't find touch very comforting, so he just turned and looked his brother in the eye.

"We're going to figure it out, buddy," Zach said firmly. "You're the smartest kid I know, Pearl's mind is constantly analyzing, Saffi and Xi are so creative, and I'm not half bad myself." This comment made Asher giggle. "And having each of us together makes us even stronger than we would be alone. Between all of us, I don't see how we *couldn't* come up with an answer.

Chapter 17 – Saffiyah

Saffiyah just couldn't seem to come up with an answer. It was really eating away at her. The harder she thought, the more her stomach tied itself into knots. They all figured out the perfect recipe, but they didn't have anywhere to make the record attempt. Just as bad, if they didn't raise enough money to pay for a priority submission, she might not even be here for the record. The whole thing they'd worked so hard for could all fall apart. She felt helpless.

Her mind kept picturing how upset Asher looked when they couldn't come up with a plan. *This record is everything to him*, she thought. They'd all gotten his hopes up and now nothing was going to come of it. She felt awful.

She couldn't help feeling like this was her fault. If it didn't happen because of her, she thought she'd never live it down.

While lying on her bed on top of her covers, her mind spiraling, her mom knocked on the door and asked to come in.

"Hey, honey. Mind if I join you?" Saffi nodded and her mom came and joined her laying down on the bed, her hands across her stomach, both of them looking up at the ceiling that Saffi had covered in glow-in-the-dark stars. Saffi felt even more rotten; she had just gotten everything unpacked, and now they might have to pack it all away once again and move along? It just wasn't fair.

"So tell me, Saffi. What's wrong? What's going on in that head of yours?"

Saffiyah thought that maybe she'd lie and say she just wasn't feeling very well, but she knew her mom would see right through that. She sighed.

"Mom, you know how we wanted to make that record?"

"Of course, Saffi, it's all you guys ever talk about! I think it's great that you kids are all working together to make something really amazing."

"Well..." Saffi started. She wasn't sure how she wanted to approach this. She didn't want her mom to feel bad, because it wasn't her fault either. "We're trying to raise money so that we can pay for a priority application, so that we won't have to wait a long time for an answer. And if we can't figure it out or can't raise the money, we might not be able to do it for months. And by then we might..."

"Have moved," her mom finished.

"Yeah," Saffi said softly.

Saffi's mom put her arm around Saffi pulling her in close. "I'm sorry, baby. I'm looking around for more clients here, so don't give up hope that we can't stay. The future is just a little uncertain right now and I just want you to be prepared. We always want to be honest with you, kiddo."

Saffiyah took a few deep breaths and tried

again. "I just...this is so important to Asher, and there's no way we're going to be able to make the record attempt if we can't get the money, and there's no way we can get a Guinness rep here this summer if we can't pay for the priority application. It would be at least 3 to 6 months, and potentially a year or more." She paused. "Mom, this is all Asher talks about. It's going to break his heart if we can't do this. I feel like everything is imploding right before my eyes and I can't stop it and my heart is racing," she said, her eyes tearing up all over again.

"Saffi, do you feel anxious?"

"Yeah, I think I do. We moved in and I suddenly had all these awesome, interesting friends from day one. I don't want to lose them, Mom."

"Oh, honey," her mom said, hugging her even tighter. "Saffi, baby, I am amazed at how resilient you have been with so many moves. It's never easy to be the new kid. I always felt that way growing up. I always looked different than all of my classmates. And it was a different time back when I was in school. Your Alchemist Club is the beginning of something wonderful, I just know it. You are all so special and amazing in your own way and I love that you all see that. What makes you different makes you stronger. We're gonna figure this out, kiddo."

Chapter 18 – Pearl

Pearl had to admit, the question of how they could organize an event in a way that could raise money for an application really had her stumped. On top of worrying about whether or not they'd raise the money, they had to find a way to actually perform the experiment at all. How do you hold a volumetrically significant amount of slime? She didn't know which roadblock was more troublesome.

She didn't think she would end up caring so much about this project, but she had to admit that it was hard seeing Asher so deflated. Pearl tried to think about what it would be like to only care about one thing and for that thing to be taken away. Even though she didn't see the practicality in caring about only one thing, she still felt bad for the kid. And she would be lying if she said she didn't care about the project, too.

She walked into the living room where her dad was reading a book. "Hey, Dad," she said, slumping down on the couch next to him.

He put down his book quickly, sensing something was wrong. "What's up, kiddo?"

"Saffi might be *moving*," Pearl said, completely deflated.

"Oh, honey, I'm so sorry," her dad said empathetically. "I know you really liked her."

"Dad, it's not just that," she said, turning to

him. "It's that everything was fitting together. Our group had the perfect vibe, we had a plan for a record, we were all working towards something, and we had our club. Without Saffi, the equation doesn't work. It doesn't fit. It all goes away," she said, balling her fists up beside her. When she got frustrated, she had difficulty slowing down enough to properly explain herself.

"Can't you do the experiment before she leaves?" her dad asked.

"No, because the normal application takes *months* to get approved. The only way to beat the line is by doing a priority application, and that costs *$800*. And we don't know how to get the money. We want to raise money with the event, but we don't know why people would care about it."

"Well," her dad thought aloud, "Why do you care about it?"

Pearl paused. She hadn't sat down and really thought about it before. "I guess...I guess I like the science. Learning the science behind it was really fun."

"Well, how can you make the science the focus of the event?"

Pearl sat quietly. Suddenly, an idea caught hold of her. She bolted out of her seat and said, "I've got to go figure things out," and sprinted into her room to brainstorm.

"Hey, Asher!" Pearl called, knocking furiously at his front door.

A groggy Asher appeared after a few moments, yawning. "Pearl, it's 6:30am. In the *summer*. I'm missing some prime REM cycles. What's so important that you couldn't wait until a reasonable time?"

"Asher," Pearl said, smiling, "I have the solution."

Asher's face lit up in shock and excitement, all signs of sleepiness gone. "What is it, what is it, what is it?" he asked, jumping up and down.

"I'll explain everything to the group later when we meet, but I wanted to tell you first. Because I know this means a lot to you, or...whatever."

"Of course it does! It's the most amazing thing I've ever done!" Asher said. "I'm sorry. *We've* ever done. I'm...I'm happy you're a part of this too, Pearl."

"Thanks, Asher."

"Guys," Asher began, "I think you're going to like what Pearl has to say."

"I have the solution — or a part of one, anyway," she began. "What we love most about this whole thing is the science, right? So what if instead of selling slime, we sell the experience of the science itself?"

"Interesting," Xi mused. "How would we do that?"

73

"Well, think about it. The reason we're interested in this isn't just because slime is fun and cool. We really enjoyed learning about the science behind it. What if we made some sort of kit or some sort of experience so that kids just like us could learn the science behind it too? What if...kids could buy supplies and instructions and packets from us with all the science in it and they came to the record breaking attempt and they made the slime along with us!

"So we set up the tables with all the individual stations. Everyone pays for the kit, and we know in advance how many stations we'll need. Then, we have a PA system, and while they make the slime step-by-step, we explain the science of what's happening. We get the kids interested in slime to learn more about science, and we get the science-y kids to have some fun with slime. We've now got a reason to get people to sign up! What if the record wasn't volume? What if the record was the number of people *making* slime at once! And then they get to walk away with a piece of the record! How cool is that?"

"Isn't that awesome?" Asher shouted, jumping up and punching the air. "Who *wouldn't* want to be a part of this? You get to make a record, and then keep a part of history!"

"Pearl, that's such a good idea!" Xi exclaimed.

"So how do we get the idea out to people?"

Xi asked.

"We raise awareness online! We could put all this on a crowdfunding site so people can pay for their spot in advance. That way we get the money beforehand, and we can pay for the application *and* pay for the supplies," Saffi explained.

"We could add more levels of perks," Zach suggested, "So if someone can't make it to the event or they just want some slime shipped to them, they can order some from the batch we make ourselves."

"We make videos, we reach out to news channels, we make posters, we go door to door, we tell everyone in any way we can!" Asher said, jumping up and down. "This idea is so cool that they would *have* to sign up!"

"Great idea," Xi agreed. "And Saffi, you can add in the description that we're raising money for charity. That always inspires people to donate more!"

"What do we want to raise money for?" Asher asked.

"Well, we all love science, but not everyone gets the same opportunities we do," Saffi started. "When I lived in Brooklyn, I had some friends in other schools that didn't really get the chance to go on field trips or do many experiments in the classroom because the school didn't have the budget for it. Maybe we find a charity that raises money so that schools that don't have as much funding can support more science clubs and opportunities?" Saffi offered.

Zach put his hand on Saffi's shoulder supportively. "That's a great idea."

"Where would we have it? Where would the event happen? What place could hold such a big event?" Pearl wondered.

"I think I know just the spot," Saffi said.

Chapter 19 – Saffiyah

Saffiyah looked around the room. *Take a deep breath*, she reminded herself.

"Okay, everyone. We've got a lot to do and don't know how much time we have to do it. If we want to raise the money," she went on, "then we have to make sure we get the word out. That means a social media campaign, crowdfunding page, and eventually getting media coverage. This means we have to make all of the digital content that goes up there. Do any of you know how to take good pictures and make videos?"

"I'm pretty good at taking cool pictures," Xi offered.

"And I actually really like vlogging," Zach added. "So I know my way around editing software."

"That's awesome. Because we can study how to build a campaign all we want, but if the content we're sharing it's quality, then it won't go anywhere," Saffi explained. She went on, "One of my mom's friends is in marketing. I spent some time with her — her name's Miranda — last night crafting a targeted marketing strategy both for coverage of our record attempt and for our slime sales afterward. Of course, for this to work, I'm going to need each of you to help. So far so good?"

She looked around the room at the smiling, nodding faces of her friends and felt a smile grow on her own face as well.

She looked at her list: social media, website, charity outreach, media pitches. Yeah, it was all there. The part of carrying it all out? That was another matter altogether.

She took another breath to calm herself and dove in head first.

"Pearl, I know you have some tech-savvy friends. Can I put you in charge of designing a Kickstarter? This is definitely somewhere where I need your help to ensure this gets picked up beyond where we're able to pitch it. If we don't make our goal, all the money will be returned to the backers, so it's all or nothing."

Pearl simply nodded as Saffiyah handed her a page of notes. "Got it."

"Zach, Asher, can you two work together on people in our area with an already established social media presence to post about the record attempt? Asher, I know you know the community of world record folks and the hashtags to work with there. Zach, I know you have some friends with successful YouTube channels and Instagram accounts in our area. I want the two of you to work with these notes Miranda and I drew up." She handed them a packet. "Let me know what questions you have after you look through this.

"Pearl, after we get the KickStarter up, let me know and get me the login so I can do all the writing.

"Before any of that, we've all got to work together to get all the photos and videos to make our crowdsourcing page look great."

"Photo shoot time!" Asher said, putting up jazz hands.

Chapter 20 – Zach

Zach was impressed. When Saffiyah assigned all the tasks to them, he'd thoroughly thought they were tackling an impossible job — despite the confidence and efficiency with which she had assigned those tasks. No matter how airtight their plan was, how were he and Asher supposed to get celebrities to care about their world record attempt?

And yet, they'd done just that.

With all the presales pouring in, they raised enough money to put in a priority application, they got approved, they had a date, and the Liberty Science Center volunteered to let them host the event in their parking lot, just like Saffi envisioned.

Asher did a fantastic job of convincing several of the world record social media accounts to feature their attempt once they had a date picked, and the orders for kits just kept rolling in on the Kickstarter as the date got closer and closer.

Above and beyond that, Saffi booked local news outlets and radio stations to cover the day of the event, and the coverage on some of those social media accounts had gone viral.

"You'll never believe the record these kids are trying to set," read one headline Zach had seen shared on Facebook. Other headlines were similar: focusing on the record, but also on how impressive it was that these kids were pulling it off.

Zach never really stopped to think about it before. They had just *done* it; they never really stopped to think that they couldn't do it because they were kids. Even a bunch of ragtag kids could do the impossible. Regardless of any additional hurdles they faced on a day-to-day basis, they all felt just as capable as anyone else.

As a result of the upcoming record attempt, the town was in a frenzy. News crews kept trying to interview their parents, their teachers, their friends. There was never anywhere to park; Zach had never seen so many people around town on bicycles. But, as he saw one resident mention on the news, "it's faster to bike for groceries than circle forever looking for a parking spot."

And the record attempt wasn't even scheduled until tomorrow. The group was hiding out in Saffiyah's basement as a way of staying out of the spotlight— well, mostly, anyway. Xi was giving interview after interview, which, the rest of the agreed, was definitely the best option. Xi was confident, funny, and gave a great interview. "And," as Pearl noted, "if they're on TV, I don't have to be!"

The rest of the crew was going through last-minute assignments and making sure everyone was clear on what they were each expected to do tomorrow on the record attempt day.

As Zach listened to Saffiyah run through

their assignments — as well as which parents were taking care of which additional pieces, stuff like working with the parking volunteers — he couldn't help but smile.

He loved his brother Asher. He loved how much Asher loved world records. And he loved how much their group of friends loved being there for Asher. Still, never in his wildest dreams could he have anticipated the group coming together the way they had in forming the Alchemist Kids Club. He never could have imagined that they would learn so much and share their passion for science with kids just like them so more people could enjoy the science they loved.

And yet that's exactly what they were about to do. Tomorrow they were going to set a world record for the largest crowd of people making slime simultaneously. They were going to teach so many people about the science of boron, get it confirmed as a world record by an official Guinness Book of World Records adjudicator, and donate all the profit to a charity that helps students in Brooklyn improve their science programs.

It was wild to think about. Absolutely mind-blowing. And yet here they were.

"Zach?"

He realized Saffiyah was looking at him. "Yeah?"

"I was just confirming that you and Asher have everything you need for tomorrow and that your parents will drive the truck of supplies to the parking lot in the morning."

"Oh. Yeah. Yeah, definitely." He looked at his brother. "Right?"

"Yeah, we have everything we need for tomorrow, Saffi," Asher chimed in. He was beaming, absolutely glowing. Happy as a clam, he thought—and then he laughed a little to himself, thinking about how much Asher hated that phrase.

If he's this happy today, Zach thought, *I can't wait to see how happy he is tomorrow!*

Chapter 21 — Asher

Today's the day, Asher thought. *Today is the day, the day everything changes.*

Asher didn't think he'd slept more than a few minutes all night — and he still wasn't tired. He just kept tossing and turning, unable to stop feeling so excited.

It was *finally* record day!

They'd been working toward this for what felt like forever, but at the same time, it felt like it all came together so quickly. Between the idea, the experiments, the research, and even the time when it looked like they weren't even going to get the chance to try to set the record at all...it'd been a busy few weeks, especially the mad rush of the last ten days once they found out they had the go-ahead from Guinness.

Asher bounded across the hall. "Zach! Zach! Time to get up!" He could hear his brother laughing in response.

"I know, bud, I'm already awake!"

Downstairs, Asher grabbed himself breakfast and reviewed his to-do list for the day — not that he needed to, of course. He'd memorized each step of their plan: even the parts that weren't on his personal to-do list.

"Asher, slow down!" his mom implored, laughing. "I know you're excited..."

"So excited!" Asher shouted.

"But you can't set a record if you make yourself sick," she finished. Asher tried to slow down, but he almost couldn't help it.

As soon as he finished eating, he headed outside to double-check that everything was in its place in the supply truck his parents had rented to haul all the supplies to the Liberty Science Center.

Before he knew it, he, his parents, and Zach loaded themselves into the truck, drove to the parking lot, and started unloading the supplies. He looked up in awe of the Liberty Science Center. He couldn't believe they were going to set a record in front of one of his favorite places on earth. It didn't seem real.

They were ahead of the schedule Saffiyah had written up for everyone, and still there were people milling about, already trying to get a good spot to see the record attempt.

"Excuse me." A tall man in a suede jacket hijacked Asher's attention from the boxes he was carrying. "Can you tell me where I might find the adult in charge?"

Asher wasn't sure what to say. Saffiyah had done the prep work, after all; there really wasn't an adult in charge so much as a fair number of adults helping. Saffiyah was really the one in charge of this operation; even as much as this was Asher's dream, he knew he wasn't the one in charge.

"I'm looking for a woman named Saffiyah," the man clarified. "Could you tell me where I might find her?"

"She's over there." Asher pointed to Saffiyah. He saw the man's face scrunch up.

"I think you misunderstand me," he said. "I'm looking for the adult in charge. Her name is Saffiyah."

"Yeah." Asher looked at the man, wondering how to make it more clear. "That's Saffiyah. She's the one that filled out all the paperwork, and is the one coordinating this record attempt. She's in charge."

Asher looked over at her and saw that she was looking his way. "Hey, Saffiyah! Can you come here? This man is looking for the woman in charge," he told her when she came closer. "I told him that was you."

"Hi!" She began brightly, "I'm Saffiyah, and I'm running as point-person today. What can I help you with?"

"Oh, um. Well, I'm Mr. Wright..."

"Oh, the adjudicator! Nice to meet you. Yes, I'm the Saffiyah you've been getting emails from."

To Asher's amusement, he watched as the man flustered his way through an explanation. This man in front of them, the man unable to believe that Saffiyah was actually in charge? He was the Guinness Book of World Records adjudicator, the man responsible for verifying whether or not they actually set a world record.

Asher listened in as the man explained to Saffiyah the procedures he was responsible for following as he verified their attempt. It was nothing new to Asher — he'd known how the adjudication process worked for years and had explained the process to his friends enough times that they knew it nearly as well as he did.

"Hey, just a moment." Saffiyah turned from the adjudicator to Asher. "Asher, aren't you supposed to be helping Pearl double-check the number of kits at each table?"

Realizing she was right, he left Saffiyah with the adjudicator to find Pearl. They had to stay on schedule to make sure everything was perfect for record day!

Chapter 22 - Saffiyah

Saffiyah couldn't help but laugh to herself at the adjudicator's reaction. Of course he couldn't believe she was in charge! Who could believe a group of kids put this whole event together?

She was glad to see that he'd come around, though; when he'd seen her clipboard with the timetable, checklist, and roles and responsibilities sheet, he understood just how organized they truly were.

Who says kids can't get stuff done?

Once Mr. Wright had gotten all he needed from Saffiyah, she moved along to make sure that the parents knew how and where to set up the tables. They needed to make enough space for the people who were going to be making the record-breaking slime along with them.

"So how do you want the tables set up, Saffi?" Pearl's dad asked as she approached him. He was extremely excited and ready to help.

"So we're thinking of making big "U" shapes, the U getting larger and larger as you get further from the center "U." We'll be set up at the front of the U on the platform, so everyone can see us," Saffi explained. "Just like on this concept chart I've got here," she said, pulling out the proper format from her clipboard.

He examined it closely and said, "We can

certainly do that. And you want the prepaid packets all set up on the tables?"

"Exactly. And they're already labeled with the name of the participant, so if we could set them up three per table, keeping them in alphabetical order, that would be great. I like to stay organized," Saffi explained.

"Right," he agreed. "You got it. And Saffiyah," he said, calling out to her as she started to march off to her next task.

"Yes?"

"I..." he began. "I just wanted to say that I really, really appreciate you all for including Pearl in this. I know sometimes she has difficulty making friends and this has been such an amazing experience for her."

"You don't need to thank us," she explained. "There's no way we could have imagined doing this without Pearl! She's an integral part of the team. We need her."

Chapter 23 - Xi

"Alright, alright, future scientists of America, it is time to begin," Xi began, speaking into the headset mic they had on their head. They were designated as the MC of the day, since they were comfortable in front of a crowd and fun to listen to. They certainly didn't mind being the center of attention.

"Today, you are all becoming a part of something bigger than yourselves: a part of a world record!" Xi paused to allow for an explosion of cheers.

They could get used to this.

"Before each and every one of you is your bowl, ingredients, and information sheet with all the written instructions for those of you who are a part of the hard-of-hearing or deaf community. Science is for everyone, so if you need additional assistance, just flag down one of our volunteers and we'll help you make some slime!" Xi gave a nod to Saffi's dad who approached a young girl who waved her arm to ask for help. Her wheelchair was just a little too short for her to properly reach the ingredients, so Saffiyah's dad could help her reach the ingredients and set the bowl on her lap.

"Are we ready, team?" Xi called out, looking behind them to see Saffiyah, Zach, Asher, and Pearl in position at the tables beside them.

They each stuck out a confident "thumbs up"

to show they were ready.

"Alright! Let's begin, people!" Xi called out to another round of claps and cheers. "So. Slime. We all know it. We all love it. That's why we're here! But slime is more than just a super fun thing to play with or throw at your brother."

Asher and Zach laughed aloud.

"Slime is what's called a non-Newtonian Fluid. It's because it flows like a liquid, but its flow changes based on stress and pressure. The viscosity of water doesn't change based on how much you squeeze it. But slime does.

"So what's viscosity? It's how much a liquid resists its flow. So honey is a viscous substance, because it takes forever to pour out of a bottle, but water has a low viscosity, because it's super easy to pour all over yourself.

"Newtonian fluids are your normal liquids: water, oil, Gatorade. They act like the normal liquids you know and love. Isaac Newton studied these fluids, and named them after himself. I mean, if I spent all that time in a lab looking at water, I'd name theories after myself, too!

"This is where slime comes in. Slime is weird. Squeezing, pulling, stirring, and playing with a non-Newtonian fluid changes the viscosity. The more you play with it and handle it, the more viscous it becomes, so much so that you might think it's a solid. It oozes like a liquid, but you can pick it up and handle it like a solid. But here's the crazy thing: non-Newtonians aren't solids or liquids; they're their own thing all together. It has the best of both liquid and solid states. When's the last time you tried to pick up a handful of water and pass it to your friend? I bet it didn't go so well! Just with your hands, you can change the state slime appears to be. How wild is that? So let's get started and make some science happen!" The crowd erupted into cheers.

"First! Squeeze your bottle of glue into your bowls," they instructed. Asher switched the slide on the screen to show facts about glue.

"Why glue, you may ask?" Xi started, "Well, glue is important to slime because slime is a polymer. Glue is considered a liquid, because these polymers can slide over each other, which makes the glue flow. It's very viscous, so it flows slowly. This polymer plays an extremely important role later when we add the most *important* ingredient." Xi waited for everyone to finish up adding their glue. Some people had more trouble getting it out of the bottle than others.

"Alright, we are ready to move along to the next step! Add your shaving foam!" They looked behind them to make sure the team was keeping up.

"Shaving foam is an important ingredient for fluffy slime, which is what we're making today,"

Xi explained. "And we're making fluffy slime because it's light but expands to a greater volume. Who can tell me what volume is?"

Chapter 24 - Zach

Zach couldn't believe how well this was going. He was looking out at the crowd of kids working on the slime and beamed with pride when he saw how many smiles there were out there. People were enjoying this! They were learning about all the scientific properties that the Alchemist Kids Club knew and loved. Who knows? *Maybe someone in this crowd would fall in love with science and make an amazing discovery in the future.* Just the thought gave him goosebumps.

"Alright, and now let's add some saline sol-u-tion!" Xi called out, drawing out the word like a sports commentator. "And what do a bunch of people use saline solution for?" Kids shouted out the answer throughout the crowd.

"That's right! Contact solution!"

Zach changed the slide and gave Xi the thumbs up.

"Alright? Are you all ready for this? It's the most important ingredient of all! Without this ingredient, the chemical reaction doesn't occur, and we just have a pile of wet glue."

Chapter 25 – Pearl

"It's time for Borax, people!" Xi shouted. As they began to explain the scientific properties of boron, Pearl started mouthing along the chemical formula. She couldn't help it. She was so excited.

"So, glue by itself has those big molecules that make it viscous, since the molecules have trouble sliding past each other. Once we add the element boron, the borate ions link with those molecules, making the sliding even more difficult. It's just one big chemical mess that gives you the slime we all know and love. So let's get mixing, people!"

Pearl's friend Hazel had even started airing a live stream on the Alchemist Kids Club Facebook page to the thousands of followers they'd picked up in the six days since the page launched. Additionally, the local Channel 12 TV station was live on site getting footage of all the kids at the different stations working together to make their own slime and record.

"Let's hold up your slime, people! Show us what you got!"

Everyone lifted up their slime, pulling it apart, squishing it in their hands, and smiling as they played with their own creations. Pearl saw tears forming in Asher's eyes.

Xi took off the headset mic and started to

speak into it while holding the headstrap. "Alright, Mr. World Record," they said, pointing at the adjudicator, "We had 637 people sign up for this event, with fifty extra kits for any walk-ins. Can you give us an official headcount?"

The adjudicator walked forward and spoke into the microphone. "Yes. So 634 people signed in, and 36 people requested a walk-in kit. With you four making your own slime, that's a total of 674 participants. The headcount confirms this number."

Xi turned back to the crowd. "And that," they paused for dramatic effect, "is a new world record for the largest crowd of people making slime at once!"

The crowd erupted in applause. Pearl couldn't believe it: they had done it.

"Hey, Asher, you okay?" Pearl said quietly, leaning in close to Asher so he could hear her over the crowd.

Asher blinked and a tear formed in his eye. "I'm just so happy."

Chapter 26 - Asher

It'd been a couple of weeks since they'd set the record, packaged all the slime, filled all the orders, and finished all the interviews. Even now, Asher couldn't quite believe it had actually happened: he was a world record holder. Every morning, he'd wake up and pinch himself; it really happened, but it still didn't feel real.

It had seemingly taken each of the kids a full week just to recover from their physical and social exhaustion after the record attempt. It turned out they couldn't find a charity that exactly fit what they were looking for, so they decided to start one of their own. With the $3,170 they made from the sales after expenses and the matching donations pouring in, they were planning on setting up a robotics club at Saffi's friends' school. They were still working out the details, but couldn't be more excited.

The best news of all was that Saffiyah's mom had found a new client. Saffiyah was staying! As soon as they knew she was sticking around, they were already trying to decide what their next project should be for the Alchemist Kids Club, but until they figured that out, Xi had made the well-made point that they deserved a chance to relax and celebrate.

Saffiyah's mom mentioned that she could get

them an invite to a special tour of the Statue of Liberty and Ellis Island. The group immediately agreed that it sounded like a fun trip, especially after Pearl admitted she'd never been.

"I've always wanted to see Ellis Island and the Statue of Liberty," she said, "but my dad and I have just never made the trip."

Saffiyah's mom had even managed to hook them up with special passes. "Look at how official we look," Zach said, "we're VIPs!"

Asher had to agree. And it was pretty cool being an important person. Their passes meant they didn't have to wait in lines, and there was a tour guide waiting for them when they got to Ellis Island.

The tour guide led them through the museum, noting just how many people came through Ellis Island each year. They looked at some of the old documents, and learned how people's names often changed upon arriving at Ellis Island. If the immigration agents decided their name was too difficult to pronounce or spell, they just wrote whatever they wanted. Sometimes immigrants themselves changed it when asked because they wanted to fit in better in their new country. And, they learned, as many as 2% of the people who arrived at Ellis Island were turned away for not having enough money, being deemed unskilled as workers, or having mental health issues. Those were the official reasons, the docent said;

"Sometimes people were sent away," she said, "simply because the immigration officer didn't like them." For that reason, Ellis Island was sometimes called Heartbreak Island.

Asher had to admit he learned a few new things every time he visited Ellis Island. The history was fascinating, and it was also interesting to him to see how immigrants weren't often treated any better then than they were now; they were considered outsiders, told they didn't belong, and frequently sent away for not very good reasons. And quite a few scientists had come through Ellis Island, including aviation pioneer Igor Sikorsky, cosmetologist Max Factor, and even science fiction writer Isaac Asimov.

As the guide explained some of the marks immigration officers and medical staff used to denote individuals they didn't think should qualify for immigration, Asher realized most of his group wouldn't have been allowed into the country. In fact, he thought, he wasn't sure any of them would have been allowed in.

Saffiyah was black: racist immigration officers might have rejected her on that fact alone. Pearl might have been allowed in, but her tendency to bristle against authority likely would have peeved someone important off. Xi's androgynous identity would have immediately disqualified them, as would have Zach's half of a right arm. Asher was pretty

101

sure that he would have been labeled different, too. Because he was on the spectrum, being in a room that crowded with that much uncertainty could have made him panic. The officers wouldn't care how much he knew about science; all they would have cared about was if he seemed "normal" or not.

But what even is normal?

Looking around the group, he couldn't help but feel thankful that they were all growing up in the community where they were, and didn't have to pass any tests to live there.

Ellis Island, he thought, was a good precursor to the Statue of Liberty, housed just a short ferry ride away on Liberty Island.

America was a country of symbols, he thought. The giant copper statue— 151 feet tall, or over 300 feet tall when including the base pedestal— was one of the first things many of them saw when arriving in their new home. It was a representation of future possibilities, even if their treatment at Ellis Island didn't always live up to that ideal.

Their record attempt had started as an idea. The new lives so many immigrants had found in the United States started as an idea. The Statue of Liberty started as an idea — as the Frenchman Édouard René de Laboulaye had noted, the United States in part owed its independence to French help, and any monument to that independence should reflect that cooperative spirit— and became a reality thanks to the work of the French government and sculptor Frédéric Auguste Bartholdi.

It was hard to imagine such a gift today, or such a majestic collaborative effort. The pedestal, for instance, almost wasn't built. But Joseph Pulitzer—after whom the Pulitzer Prize is now named—led a fundraising drive to help make it happen. The sculpture itself took decades to come together, from its proposal in 1865 or 1870 to its dedication in 1886.

And copper, he thought, was a symbolic choice. Sure, the oxidation gave the statue a dignified greenish hue, but more than that, he liked that copper is a malleable and highly conductive metal.

He turned to Pearl, "You ever think about how copper was a perfect choice for the statue?"

She looked at him questioningly. "What do you mean?"

"I mean, copper is conductive. It carries energy without losing very much of that energy. It feels like that should be true of any great idea like liberty, you know?"

She nodded. "And copper's malleability is pretty cool, too."

He smiled. It felt good to have a friend who understood the way he thought.

Printed in the USA
CPSIA information can be obtained
at www.ICGtesting.com
BVHW050053250823
668863BV00004B/30

9 798218 008680